Biology
Required Practicals
Exam Practice Workbook

Primrose Kitten

Contents

Introduction

As part of your AQA GCSE Biology course, you will carry out ten Required Practicals. You can be asked about any aspect of any of these during the exams; this can include planning an investigation, making predictions, taking readings from equipment, analysing results, identifying patterns, drawing graphs, or suggesting improvements to the method. You can also be asked about practicals that are similar but that you may not have done before. You need to be able to recognise and apply the key practical skills that you have learnt to different experiments.

These practical questions account for at least 15% of the total marks. This Exam Practice Workbook allows you to practise answering questions on the ten Required Practicals and become familiar with the types of questions you may find in the exams. There are lots of hints and tips about what to look out for when answering practical questions.

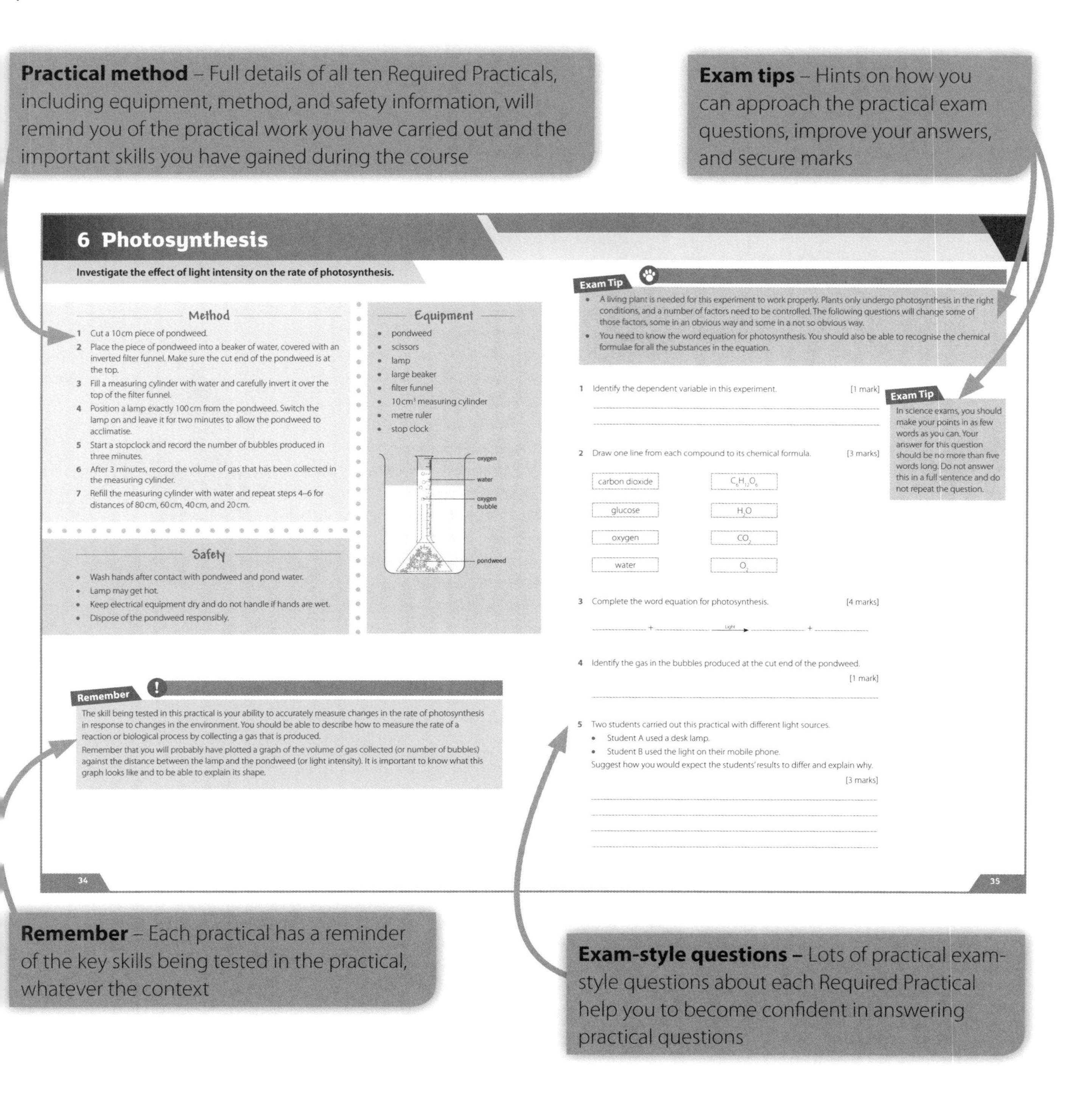

1 Microscopy

Use a light microscope to observe, draw, and label biological specimens.

Method

Before you can look at the cells on the slide, you will need to set up your microscope.

Most microscopes have a built-in light source, but if the one you are using does not then you need to arrange the mirror found underneath the stage so that light is directed through the lens system.

1. Move the stage to its lowest position.
2. Place a prepared slide on the centre of the stage and fix it in place using the clips.
3. Select the objective lens with the lowest magnification and raise the stage to its highest position.
4. Look through the eyepiece and slowly move the stage down by turning the coarse focus adjustment until the cells on the slide come into view.
5. Turn the fine focus adjustment to sharpen the focus so the cells can be clearly seen.
6. If you wish to view the object at greater magnification to see more detail, switch to a higher magnification objective lens and use the fine focus adjustment to sharpen the focus.

Safety

- Take care when handling glass slides as they are very fragile.
- Take care not to break the slide by moving the stage too close to the objective lens.

Equipment

- light microscope with low and high power objective lenses
- a range of prepared animal cells including:
  - cheek cells
  - red blood cells
- a range of prepared plant cells including:
  - onion epidermal cells
  - leaf palisade cells

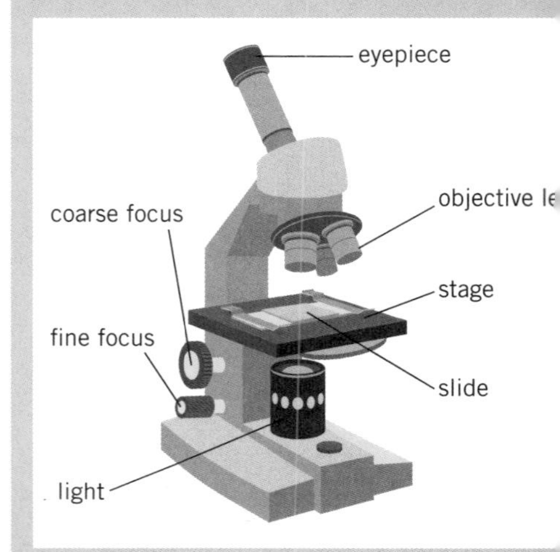

Remember

The skill being tested in this practical is whether you can use a light microscope to observe plant and animal cells. You need to be able to describe how to set up the microscope, focus on a slide containing the specimen, and then make a labelled scientific drawing of what you see. Don't forget to include the magnification in scientific drawings.

To calculate the total magnification of the microscope you used to see your cells:

total magnification = eyepiece lens magnification × objective lens magnification

Exam Tip

In the exam, you may be asked about a specimen you haven't come across before. It's important to remember that it's still just a plant or animal cell. Apply your knowledge of the microscopy practical to the exam question, whatever the specimen.

1 Label the diagram of an animal cell. [5 marks]

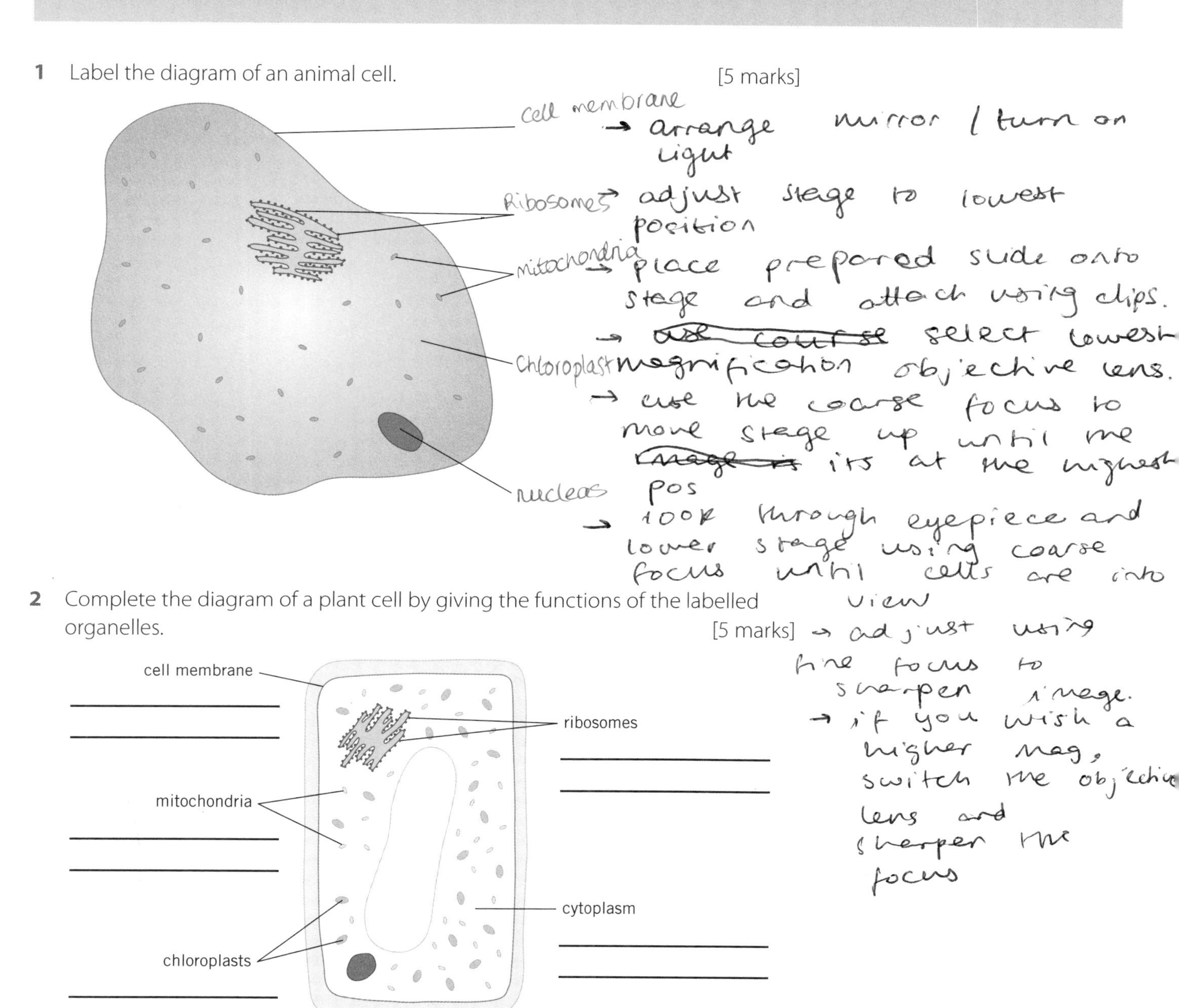

2 Complete the diagram of a plant cell by giving the functions of the labelled organelles. [5 marks]

3 A student examines an onion cell under a microscope. Suggest why this plant cell is not green. [1 mark]

It does not contain chloroplasts

4 Draw **one** line from each part of the microscope to its function. [4 marks]

Part	Function
Slide	Part that can move around, so you can view different sections of the sample
Eyepiece	Smooth curved piece of glass closest to the sample being viewed
Objective lens	Piece of glass where the sample is immobilised and stained
Stage	Part of the microscope that you look through

5 Suggest why it is not possible to see the internal structures of a bacterial cell using a light microscope. [1 mark]

It does not have a high enough magnification

6 Which objective lens should you use when you first focus a microscope on a slide? [1 mark]

Tick **one** box.

×4 ☑

×40 ☐

×100 ☐

×400 ☐

Exam Tip

For this question you need to select ONE answer.

You will not get any marks if you tick more than one box, even if one of the boxes you tick is correct.

7 Suggest why it is important to state the magnification on any drawings you make from a microscope. [1 mark]

So that the actual size can be calculated

8 A student drew three images of a plant cell at three different magnifications but failed to label the drawing with the magnification. The microscope has three objective lenses ×40, ×100, and ×400.

A

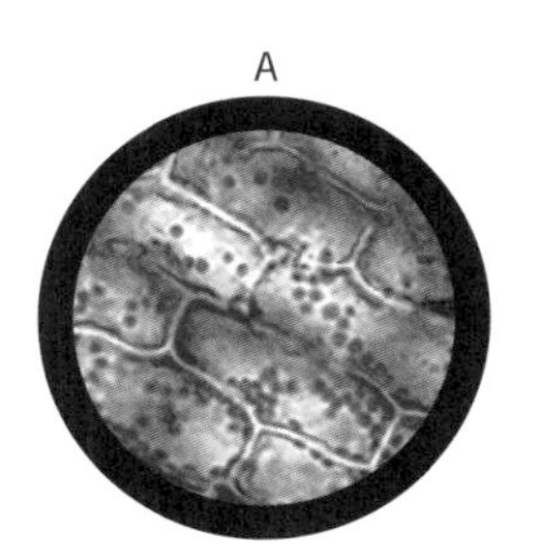

B

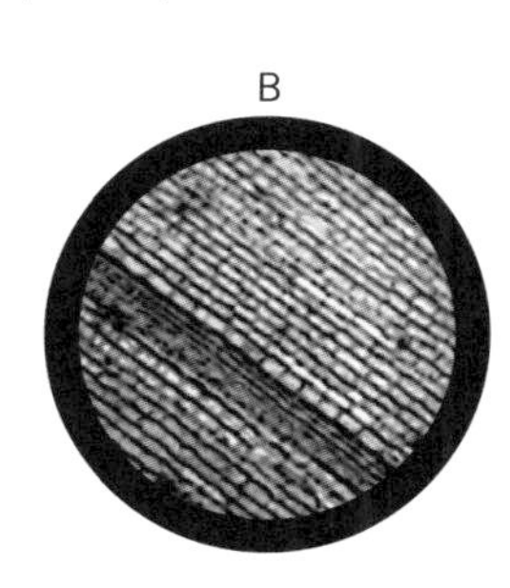

C

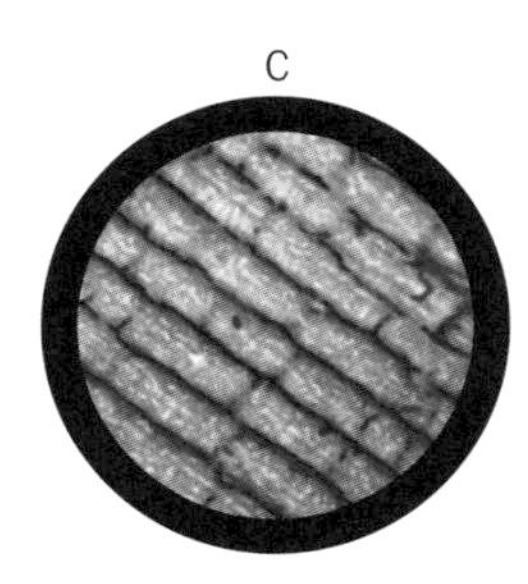

Draw **one** line from each image letter to the correct objective lens. [3 marks]

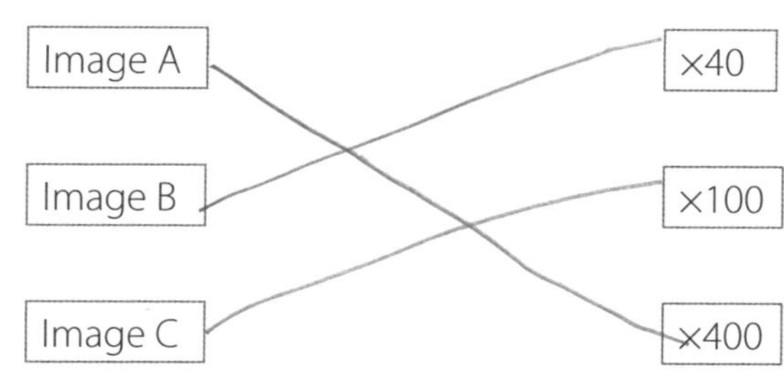

9 A student has prepared a sample on a slide and wants to view it using the ×100 objective lens.

Describe the steps the student should take to focus the microscope. [4 marks]

____

____

____

____

____

____

10 The two different classes of microscopes are light microscopes and electron microscopes.

Compare these two different classes of microscopes. [6 marks]

Electron microscopes have a higher magnification and resolution than light microscopes but are more expensive.

____

____

____

____

____

____

____

Hint

When 'compare' is the command word in a question, you need to give similarities and differences.

11 Put these objects in order of size, **smallest** first. [2 marks]

A Animal cell (10 μm)

B Ant (1 mm)

C DNA (10 nm)

D Bacterial cell (1 μm)

E Virus (100 nm)

C	E	D	A	B

Exam Tip

When converting between units, the answer is going to be a 'sensible' size of number. If the answer you get is massive or tiny, you may have multiplied numbers when you needed to divide, or divided numbers when you needed to multiply them.

12 A sample was measured to be 125 000 μm.

Calculate this size in cm. [1 mark]

Size = 12.5 cm

13 Magnification can be calculated by dividing image size by the size of the real object.

$$\text{magnification} = \frac{\text{image size}}{\text{real size}}$$

Rearrange this equation to show how the real size of the object can be calculated. [1 mark]

real size = image size x magnification

14 The image below shows pollen grains viewed under a microscope. A special slide is used with a scale printed on it.

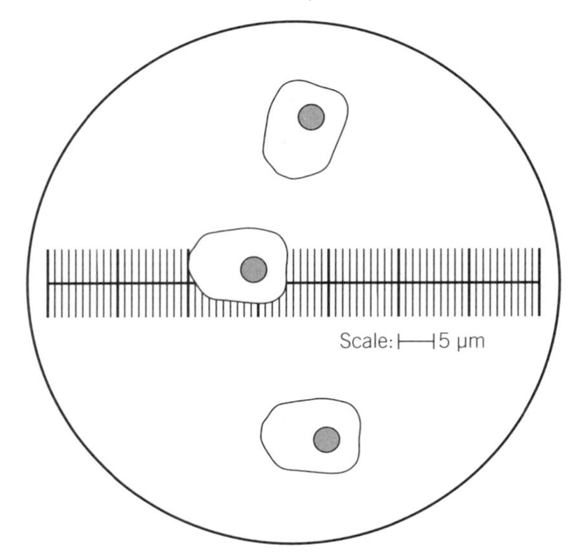

Use the scale to determine the length of a pollen grain. [1 mark]

Length of pollen grain = 14 µm

15 **H** A student views a sample using a ×4 eyepiece and a ×100 objective lens. They measure it to be 2 cm long. Calculate the real size of the sample in µm. [3 marks]

$\frac{2}{400}$ = 0.005 cm = 50 µm

Real size of sample = 50 µm

16 **H** The nucleus of an animal cell has a diameter of 6 µm and the diameter of the whole animal cell is 100 µm. You can assume they are both perfect spheres. Calculate how many times larger the volume of the cell is compared to the volume of the nucleus. Give your answer as a whole number. [5 marks]

$\frac{\frac{4}{3}\pi 100^3}{\frac{4}{3}\pi 6^3}$ = 4630

Answer = 4630 times larger

Exam Tip

You can be asked to apply maths in lots of different ways. You'll need to know equations from maths lessons in your science exam as well.

volume of a sphere $= \frac{4}{3}\pi r^3$

2 Microbiology

Use agar plates and measure the clear zones produced around colonies.

Method

1. Thoroughly clean your benchtop with disinfectant.
2. On the bottom of the nutrient agar plate, use a marker to write:
   - the date
   - your initials
   - the name of the bacteria you are using.
3. Divide the plate into three equal sections labelled A, B, and C.
4. Use antibacterial soap to wash your hands.
5. Place a different antibiotic disc in the centre of each section of the nutrient agar plate.
6. Replace the lid of the nutrient agar plate and attach it with one piece of sticky tape on each side. Do not create an airtight seal by taping the lid the whole way round the plate.
7. Incubate the plate at 25 °C for 48 hours.
8. Measure the diameter of the clear zones (zones of inhibition) around each antibiotic disc. Two measurements of the clear zone diameter should be taken at 90° to each other.
9. Record your results in a suitable table.

Safety

- Bench tops need to be wiped with disinfectant or alcohol before starting your experiment. Allow alcohol fumes to diffuse before lighting a Bunsen burner.
- Use a Bunsen burner to flame apparatus used to transfer bacteria.
- Benches must be cleaned with disinfectant or alcohol after the experiment. Naked flames must not be used during this part of the procedure.
- Equipment must be sterilised in an autoclave before and after the experiment.
- Agar plates must be incubated at a maximum temperature of 25 °C with lids attached by a few pieces of sticky tape. An air tight seal must not be formed.

Equipment

- disinfectant solution/alcohol to sterilise benches
- Bunsen burner
- sterile Petri dish containing nutrient agar
- inoculating loop
- sticky tape
- marker pen
- culture of a bacterium (such as *Bacillus subtilis*) in a small capped bottle
- tweezers
- small pieces of filter paper
- ruler
- antiseptics (e.g, antibacterial liquid soap, mouthwash, liquid toothpaste) or pre-soaked antibiotic discs

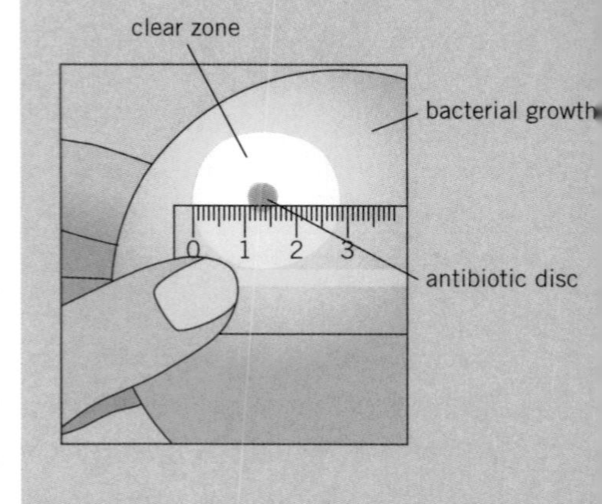

Remember

This practical tests your ability to observe biological changes and responses to environmental factors by making accurate measurements of length and area.

You should be familiar with the safety precautions required to minimise the risks of working with live bacteria.

Exam Tip

Aseptic technique is important in this practical so you should be prepared for questions to examine your knowledge of this.

As safety is really important when working with bacteria, you should make sure you know the risks associated with the practical, and any safety measures which are needed to reduce those risks.

1 Describe the measures needed to ensure this practical is carried out safely. [6 marks]

Exam Tip

When writing about practical safety or writing a risk assessment, it is important you cover WHAT can harm you, HOW it can harm you, and how you can PREVENT it from harming you.

2 Describe what an antibiotic is. [2 mark]

3 A microbiologist investigated the effectiveness of two antibacterial soaps against *Campylobacter* bacteria. The diagram below shows the nutrient agar plate after incubation.

Disc A was soaked in soap A.

Disc B was soaked in soap B.

Disc C was soaked in distilled water.

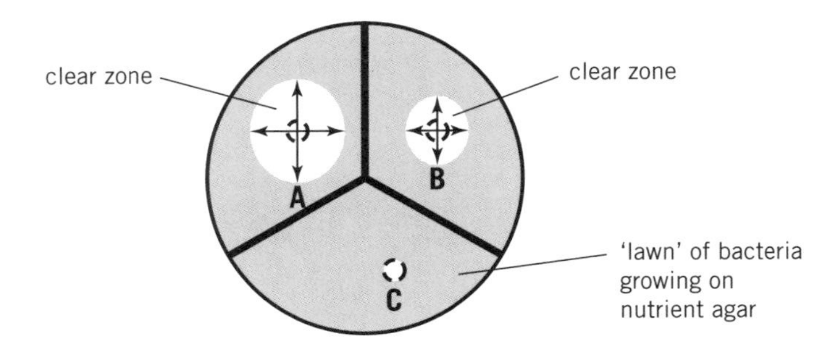

The microbiologist's results are shown in the table below.

Substance	Diameter of clear zone in mm		
	1	2	Mean
soap A	13	11	
soap B	8	6	
distilled water	none	none	

a Suggest why it is important to space out the discs on the agar. [1 mark]

______

______

b Explain why two measurements of the clear zone were made at 90° to each other. [1 mark]

______

______

c Complete the table by calculating the mean for each row. [1 mark]

d Calculate the area of the clear zone for each soap being tested. [4 marks]

Hint

area of a circle = πr^2

Make sure you use the mean values for the diameter of each clear zone.

______

______

______

______

Area of clear zone for soap A = ______ mm^3

Area of clear zone for soap B = ______ mm^3

e Identify which antibacterial soap is the most effective and explain how you know. [3 marks]

______

______

______

______

f Describe the purpose of disc C. [2 marks]

______

______

______

4 a Give **two** reasons why agar plates are usually incubated with the surface of the agar facing downwards. [2 marks]

______

______

______

______

b It is important to check whether there are any signs of contamination on agar plates when culturing *E.coli*.

Describe how you could tell if a nutrient agar plate was contaminated with other microorganisms. [1 mark]

Hint

E.coli cultures produce a pale 'lawn' of bacteria.

c Explain why you should not create an airtight seal when taping the lid to the Petri dish. [2 marks]

d Explain where on the petri dish you should write your name, the date, and the species/strain of bacteria. [2 marks]

5 a Suggest why you should never incubate agar plates at a temperature higher than 25 °C in a school laboratory. [1 mark]

Hint

There are many different strains of *E.coli* that have different optimum temperatures for growth. Some are perfectly safe but others, which normally live in our intestines, can be very harmful if ingested. Think about what the temperature in the human digestive system is.

b Describe how to use good aseptic technique when transferring bacteria from the culture bottle to a nutrient agar plate. [6 marks]

Exam Tip

Make sure you focus on what the question is really asking you about. You don't need to write a whole method here – you should focus on describing aseptic technique.

6 A student incubates bacteria in a liquid containing the nutrients the bacteria need to grow and divide.

The student then samples the liquid every hour for four hours, counts the number of bacteria in the sample, and uses this to estimate the total number of bacteria.

Their results are shown on the graph below.

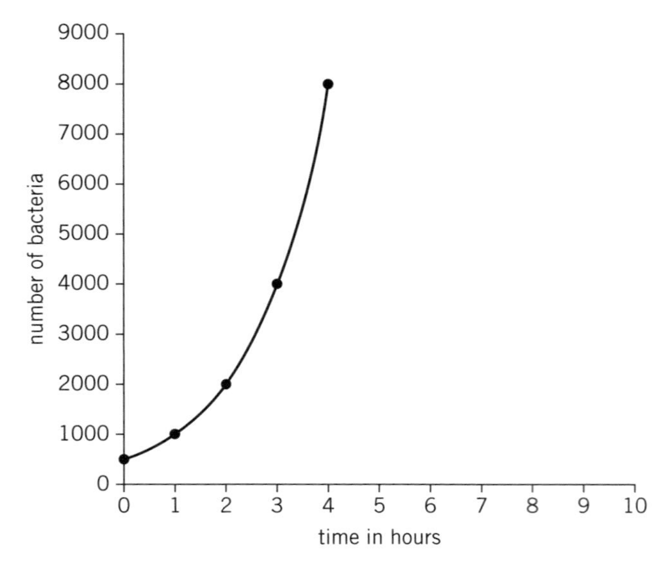

a Estimate the number of bacteria in the culture after three hours. [2 marks]

Answer = ________________ bacteria

b Suggest one way that the student could improve their method. [1 mark]

c Predict what would happen if the student continued sampling every hour for another six hours. [2 marks]

Hint

Think about the amount of nutrients in the liquid.

7 A strain of *E.coli* divides every 20 minutes.

Starting with a population of one bacterium, calculate how many bacteria there would be after 120 minutes. [2 marks]

Answer = ______________________ bacteria

8 A colony of *Mycobacterium tuberculosis* has a population of 100 bacteria.

Calculate the population size after five days if the mean division time is 800 minutes. [3 marks]

Exam Tip

You might not know the name of this bacterium but the theory is exactly the same as in the previous question. The numbers might look more complicated but the only real difference is that you need to work out the number of minutes in five days first.

Answer = ______________________ bacteria

3 Osmosis

Investigate the effect of a range of concentrations of salt or sugar solutions on the mass of plant tissue.

Method

1. Use an apple corer to cut five vegetable cylinders and cut them to the same length (making sure that no skin is left on the end of the cylinder).
2. Dry each cylinder with filter paper before carefully measuring its mass.
3. Record these values in a table as 'Mass before'.

Concentration of solution in mol/dm³	Mass before in g	Mass after in g	Percentage change in mass

4. Measure 10 dm³ of each concentration of sugar solution into clearly labelled boiling tubes (distilled water should be used in a tube labelled 0 mol/dm³).
5. Add one vegetable cylinder to each boiling tube and start the stop-clock.
6. After an exact amount of time (established by a preliminary experiment), remove the vegetable cylinders from the sugar solution.
7. Immediately dry each vegetable cylinder and re-weigh it.
8. Record these values in the table as 'Mass after'.
9. Calculate the percentage change in mass for each cylinder.

$$\% \text{ change in mass} = \frac{\text{change in mass}}{\text{initial mass}} \times 100$$

Safety

- Be very careful when using the sharp knife and the apple corer.
- Do not eat the plant/vegetable tissue samples.

Equipment

- plant tissue, e.g., potato, sweet potato, or beetroot
- range of sugar concentrations (or you could use different salt concentrations)
- distilled (pure) water
- apple corer
- sharp knife
- white tile
- filter paper
- tweezers
- boiling tubes
- measuring cylinder
- ruler
- balance
- stop-clock

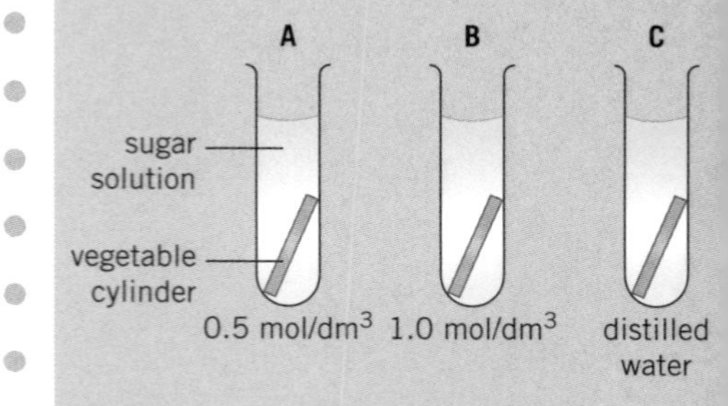

Remember !

This practical requires you to accurately measure length, mass, and volume in order to measure osmosis. You should be able to describe how to do this for a range of sample types as osmosis happens in all cells.

The easiest example to use in class is a potato because they are cheap, readily available, and easy to cut into even shapes, but the principles of this experiment can be applied to any cell. There will be a range of examples used in this section to get you used to seeing the same theory applied with different types of tissues.

Exam Tip

There are lots of opportunities for the exam board to test your maths skills in this topic. Don't get flustered by maths appearing in a biology question. Approach the maths questions the same way you would in a maths lesson.

1 Use answers from the box to complete the definition of osmosis. [1 mark]

concentrated	dilute	ions	water

The diffusion of water from a ____________________ solution to a ____________________ solution across a partially permeable membrane.

2 Give the meanings of the following keywords. [3 marks]

Hypertonic ____

____

____

Hypotonic ____

____

____

Isotonic ____

____

____

Exam Tip

There are quite a few new words in this topic. It is important to learn 'exam perfect' definitions of keywords as you will often be asked to give the meaning of them for one or two marks.

3 Identify the part of the cell that is involved in osmosis.

Tick **one** box. [1 mark]

Cell membrane ☐

Chloroplast ☐

Nucleus ☐

Ribosome ☐

4 Describe what happens to a plant cell that is placed in a hypotonic solution. [6 marks]

____

____

____

____

____

____

____

____

____

____

Exam Tip

Long answer questions are the place to show the examiner that you know lots of keywords, but make sure you use them correctly!

5 Two students carried out a version of the experiment. They were given whole potatoes and were told to core and slice them into even cylinders.

Student A insisted that they should leave the skin on the end of the potato cores.

Student B thought that they should peel the potatoes before coring them.

Explain which student is correct. [3 marks]

Correct student = ____________________

Explanation:

____

____

____

____

____

6 Identify the independent variable in the experiment described in the method. [1 mark]

Independent variable = ____________________

7 A similar experiment investigated osmosis in carrots.

Explain why the carrot cores all needed to be the same length and width. [3 marks]

____

____

____

____

8 When this experiment was carried out with samples of chicken breast instead of vegetable cores, the following data was collected.

Initial mass in g	Final mass in g	Change in mass in g	Percentage change in mass
0.92	1.35		

Calculate the percentage change in mass of the chicken breast sample. [2 marks]

____

____

____

Percentage change in mass = ____________________ %

9 The following data was collected when the experiment was performed in class.

Concentration of sugar solution in mol/dm³	1	0.75	0.5	0.25	distilled water
Percentage change in mass	−39.65	−33.54	−29.34	−21.76	+12.63

a • Plot the data on the axes provided below.
• Draw a line of best fit. [3 marks]

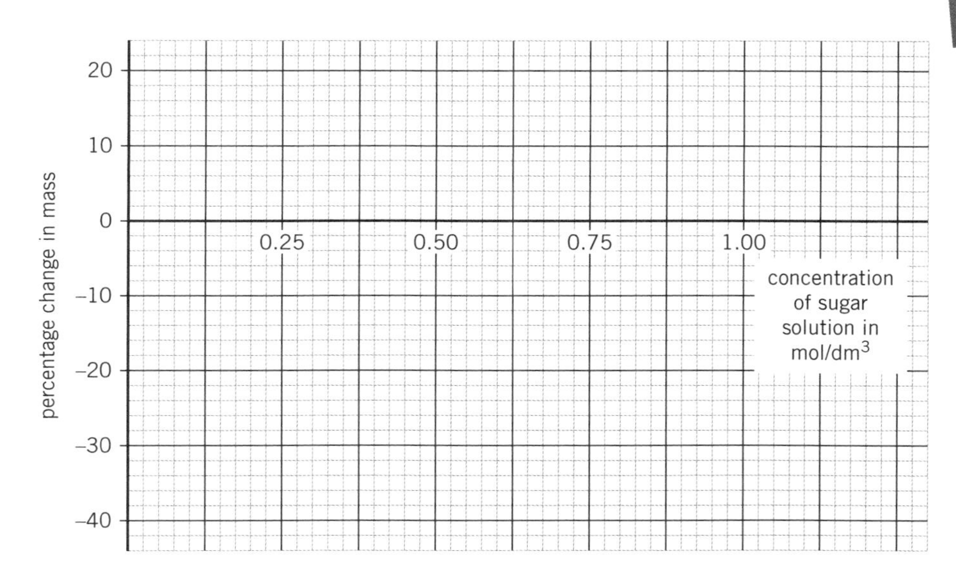

Exam Tip

Make sure you draw one smooth, continuous line of best fit. This is not art and the line should not be shaded or feathered.

b Estimate the concentration at which the sugar solution will be isotonic with the potato cells. [1 mark]

Concentration = ______________ mol/dm³

10 A teacher carried out preliminary experiments to find the best values for some of the control variables in this experiment.

- Identify two control variables that the teacher might want to determine in the preliminary experiment.
- Suggest a reason why it is important to use a suitable value for each of the variables you have chosen. [4 marks]

Control variable 1: ______________

Reason: ______________

____

____

Control variable 2: ______________

Reason: ______________

____

____

Hint

A preliminary experiment is carried out to a make sure the practical can be carried out within the time available. Think about what factors might have made the practical take too long.

11 A student placed an egg in vinegar to dissolve the shell, leaving the membrane exposed. The egg was then:

- dried, weighed, and placed it in 5% salt solution for 30 minutes
- removed from the solution, dried, and re-weighed.

The weight of the egg had not changed.

The student measured carefully and did not make any mistakes.

Explain what conclusion the student can draw from this experiment. [2 marks]

____

____

____

____

12 Two groups of students carried out a version of the experiment in exactly the same way but produced very different results.

Evaluate the methods the two groups used. [6 marks]

Group A:

- used one potato for the whole experiment
- weighed each individual potato core before placing in solution
- timed exactly 5 minutes before removing the samples
- weighed the cores immediately after removing from the solution.

Group B:

- took each core from a different potato
- weighed one core before and used this as the 'before' weight for all samples
- didn't time the experiment
- dried the samples before weighing them.

____

____

____

____

____

____

____

____

____

____

____

____

Exam Tip

When you see the command word evaluate you need to find good things, find bad things, give your opinion, and state why you came to that opinion.

13 Explain what will happen to the volume of water inside the visking tubing bag shown in the diagram below. [3 marks]

visking tubing bag
pure water
10% sugar solution

Change in volume of water inside the tubing bag:

____

Explanation:

____

____

____

14 A student cuts a cylinder of parsnip so that it has a diameter of 1.2 cm and a height of 6.4 cm.

a Calculate the volume of the cylinder. [2 marks]

____

____

Volume of parsnip cylinder = ______________ cm^3

Hint

volume of a cylinder = $\pi r^2 h$

b (H) Calculate the surface area of the cylinder.
Give your answer to 2 decimal places. [2 marks]

____

____

____

Surface area = ______________ cm^2

Exam Tip

The surface area of a cylinder is two circles (top and bottom) and all the way around (a rectangle). Surface area of a cylinder = $2\pi rh + 2(\pi r^2)$

15 (H) When collecting the sample of sugar solution needed for the experiment, it was noticed that there were lumps of sugar at the bottom of the flask.

Suggest what has happened and what affect this will have on the results of your experiment. [3 marks]

____

____

____

____

____

____

4 Food tests

Test food samples for a range of carbohydrates, lipids, and proteins.

Method

A Test for starch

1. Add a few drops of iodine solution to the food on the spotting tile.
2. Yellow–red iodine solution turns blue–black if starch is present.

B Test for sugar

1. Place a small amount of food in a test tube.
2. Add enough Benedict's solution to cover the food.
3. Place the test tube in a warm water bath for 10 minutes.
4. Blue Benedict's solution turns brick red on heating if a sugar such as glucose is present.

C Test for lipids (fat)

1. Place a small amount of food into a test tube.
2. Add a few drops of ethanol to the test tube.
3. Shake the test tube and leave for one minute.
4. Pour the solution into a test tube of water.
5. Ethanol added to a solution gives a cloudy white layer if a lipid is present.

D Test for protein

1. Place a small amount of food in a test tube.
2. Add 1 cm^3 of Biuret reagent. Alternatively, add 1 cm^3 of sodium hydroxide solution and then add a few drops of copper sulfate solution.
3. Blue Biuret reagent turns purple if protein is present.

Equipment

- small pieces of different foods (e.g. cheese, crisps, pasta, ham, bread, boiled sweets)
- test tubes and test-tube rack
- spotting tile
- iodine solution
- Benedict's solution
- Biuret reagent *or* dilute sodium hydroxide solution and copper sulfate solution
- disposable pipettes
- filter paper
- water bath

Safety

- Do not eat any of the food.
- Some people may have food allergies.
- Wear splash-proof eye protection.
- Biuret reagent: IRRITANT.
- Sodium hydroxide: IRRITANT.
- Ethanol: HIGHLY FLAMMABLE – keep away from naked flames.
- Iodine solution: HARMFUL – avoid contact with skin.
- Water in the water bath will be very hot.

Remember 

This practical is testing for four different compounds found in food; lipids (fats), sugars, starch (carbohydrates), and proteins. You should be able to identify and describe the correct method for each test.

This practical also tests your knowledge of how to safely use heating devices and techniques.

Exam Tip

Make sure you learn the names of all the food tests and reagents. In an exam you may give the right colour change for a test but you will not get credit for your answer if you give the wrong test name. Be careful when the names of tests are similar, such as the Biuret reagent test for proteins and Benedict's test for sugars!

1 Give **two** advantages of knowing what compounds are in a food. [2 marks]

____

____

____

____

2 Give the type of enzyme in the digestive system that is responsible for breaking down the following substances in food. [3 marks]

a Carbohydrates

____

b Lipids

____

c Proteins

____

3 Draw **one** line from each food group to the products when it is broken down. [3 marks]

Carbohydrates	Amino acids
Lipids	Fatty acids and glycerol
Proteins	Sugars

4 A sample was sent to a lab for testing, but the label fell off. Use the information given to suggest which substance the food contained. [1 mark]

Lab book entry — 2nd August

Test 1 – iodine test	result – negative
Test 2 – Biuret test	result – negative
Test 3 – lipid test	result – negative
Test 4 – Benedict's test	result – positive

Substance = ______________

5 A student tested a sample of food to see what compounds it contained. They added the following ingredients into the test tube.

- sample of food dissolved in 7.0 cm^3 distilled water
- 1.5 cm^3 of Biuret solution A
- 2.0 cm^3 of Biuret solution B.

Calculate the percentage of the final solution that was Biuret solution B.
Give your answer as a whole number. [2 marks]

____

____

____

Percentage of final volume that was Biuret solution B = ____________%

6 It is important to select the correct equipment when carrying out an experiment.

Identify which of the following measuring cylinders would be the most appropriate for measuring 1 cm^3 Biuret solution.
Give a reason for your answer.

[2 marks]

A 5 ml measuring cylinder

B 10 ml measuring cylinder

Most appropriate cylinder: ____________

Reason: ____

____

____

7 Read the following statements about the results of food tests.

A The starch test goes blue-black with a positive result.

B The lipid test goes purple if positive.

C The Biuret reagents test for proteins needs to be heated.

Tick **one** box to indicate which statements are true. [1 mark]

A, B, and C are true ☐

A only is true ☐

A and B are true ☐

B and C are true ☐

Exam Tip

There are four different mini practicals involved in this experiment. It is important that you learn and can recognise the results for each food test.

8 Safety information for the chemicals used in this practical is given below.

Chemical	Safety information
starch	can stain skin
ethanol	flammable
copper sulfate in Benedict's solution	irritant
sodium hydroxide in Biuret reagents	corrosive

Use the information in the table and your knowledge to describe the hazards, the risks associated with them, and how the risks can be managed. [6 marks]

Hint

Explain WHAT can harm you, HOW it can harm you, and how you can PREVENT it from harming you.

9 Some processed cheeses have starch added to change the texture and to make them easier to melt and grate.

Suggest how this might affect the results of food tests on these cheeses. [2 marks]

10 Coeliac disease is a condition that means a person cannot eat foods which contain the protein gluten.

Describe how you would test for the presence of gluten in bread. [3 marks]

Hint

Don't be thrown by the fact you may not know much about gluten. Look carefully at what the question tells you about gluten.

11 Ethanol and bile, in the digestive system, have the same effect on lipids.

Choose the term that describes the effect of ethanol on lipids and give the meaning of the term. [3 marks]

A Denature

B Digest

C Emulsify

D Absorb

Correct term: ____________

Meaning:

____

____

Hint

Bile has two functions. This question is not asking about bile's role in neutralising stomach acid.

12 Two students carried out Benedict's test for sugars.

Student A observed and recorded the colour change after 5 minutes.

Student B recorded data for 5 minutes using a colour probe (colorimeter) attached to a data logger. This constantly recorded the colour of the solution as a numerical value.

a Evaluate these two different methods of following the practical. [6 marks]

____

____

____

____

____

____

____

____

____

____

____

____

Exam Tip

When you see 'evaluate' as the command word in an exam question, you need to point out the good bits and the bad bits of both methods, give your opinion about which is best, and then explain why you came to that conclusion.

b Student A's observation was a qualitative result, whereas student B's results were qualitative.

Describe what the terms 'qualitative' and 'quantitative' mean. [2 marks]

Qualitative

____

____

Quantitative

____

____

13 Colorimeters can be used to measure the amount of red light absorbed by a sample with Benedict's reagent in it.

A calibration curve can be created using a set of solutions with known concentrations of glucose.

a To complete the calibration curve:

- plot the data from the table below
- draw a line of best fit. [3 marks]

% concentration of sugar solution	0	2	4	6	8	10
% absorption of red light	1.9	1.0	0.48	0.25	0.1	0.02

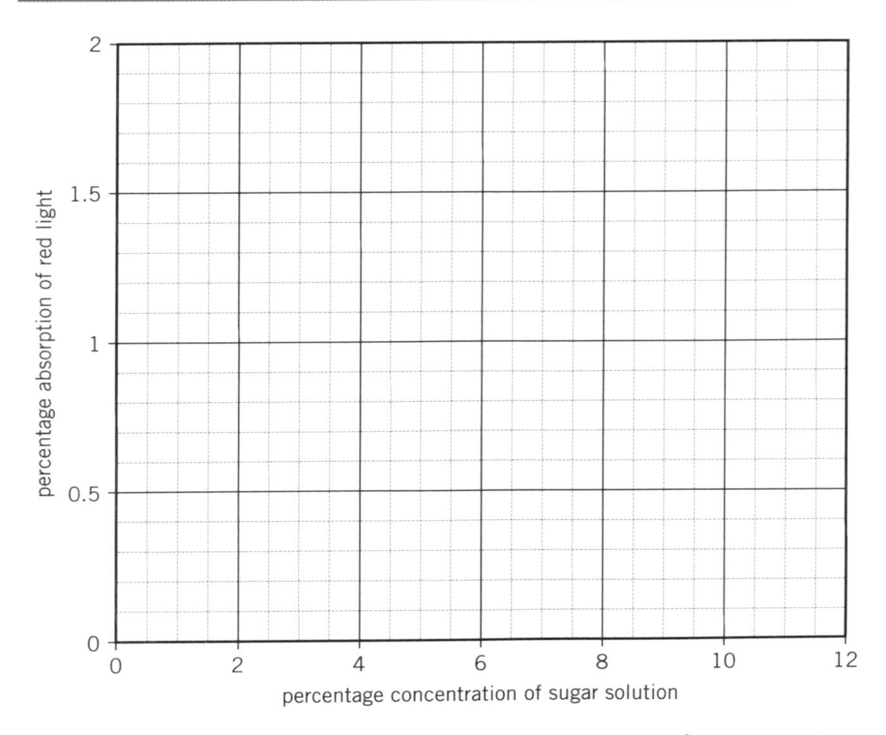

b A student tested a range of unknown samples (A–D) for sugar using Benedict's reagent and the calibrated colorimeter.

Use the student's results (in the table below) and the calibration curve to estimate the percentage of sugar in each solution. [4 marks]

Sample	% absorption of red light	Estimated % concentration of sugar solution
A	1.2	
B	0.7	
C	0.9	
D	1.5	

5 Enzymes

Investigate the effect of pH on the rate of reaction of an enzyme.

Method

1. Transfer 2 cm^3 of each pH buffer solution to separate, labelled test tubes. Use a separate syringe for each pH buffer.
2. Use another syringe to add 4 cm^3 starch solution to five test tubes.
3. Place the pH buffer test tubes, starch solution test tubes, and a test tube containing 10 cm^3 amylase solution in a 30 °C water bath.
4. Place a thermometer in one of the test tubes containing the starch solution and wait until it reaches 30 °C.
5. Whilst waiting, add a drop of iodine solution into each dimple of a spotting tile.
6. Use a glass stirring rod to transfer a drop of starch solution to the first dimple of the spotting tile. This will be your 'zero time' test.
7. When the solutions have reached 30 °C, add 2 cm^3 of the first pH buffer solution 2 cm^3 amylase solution to one of the starch solution test tubes and start a stop clock.
8. Every 10 seconds, use the stirring rod to transfer a drop of the mixed solution to the iodine solution in the next dimple on the spotting tile. Make sure the stirring rod is rinsed with water in between each sample.
9. Repeat step 8 until the iodine in the dimples does not change colour.
10. Record the time for amylase taken for amylase to completely break down the starch in a suitable results table.
11. Repeat steps 7–10 for each pH buffer solution.

Safety

- Use eye protection.
- Iodine is harmful. Avoid contact with skin.

Equipment

- amylase solution (0.5%)
- starch solution (0.5%)
- iodine solution in a dropper bottle
- buffer solutions (range of pH values)
- 5 cm^3 syringes
- pipette
- test-tube rack
- test tubes (one for each pH to be tested)
- spotting tile
- stop clock
- marker pen
- water baths set at 30 °C

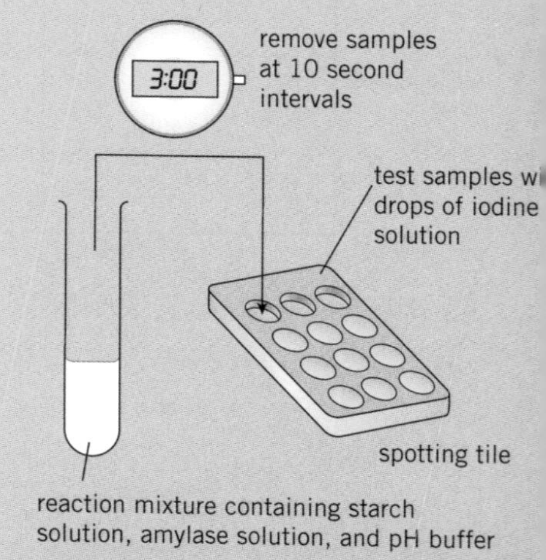

Remember

This practical tests your ability to accurately measure and record time, temperature, volume, and pH. You should know how to find the rate of a reaction by measuring the time taken for an indicator to change colour. You should be able to describe how to use a 'continuous sampling' technique to monitor a reaction. Make sure you can explain how a water bath is used to control the temperature and why this is important.

Exam Tip 

The enzyme used in this experiment is amylase, but the principles can apply to any enzyme and its substrate.

1 Define the term enzyme. [2 marks]

Exam Tip

There are lots of key words in Biology. These are often asked as one- or two-mark questions. It is worth spending the time to learn as many as you can.

2 The enzyme used in this practical is amylase.

Give the sites where amylase is produced in the body. [2 marks]

Exam Tip

Always look at the number of marks available for a question. This gives you a good indication of how many points you need to write down. This question is worth two marks, so you need to give two places where amylase is produced.

3 Describe the roles of the following solutions that are used in this practical. [3 marks]

Starch solution

Iodine solution

4 Suggest why it is important to have a drop of iodine in the spotting tile as a 'zero time'. [1 mark]

5 This experiment tests how pH affects the rate of an enzyme-controlled reaction. Identify two other factors that would affect the rate of an enzyme-controlled reaction. [2 marks]

6 Catalase is a substance found in the liver that helps to break down hydrogen peroxide into water and oxygen.

Use the 'lock and key' model to describe how catalase breaks down hydrogen peroxide into water and oxygen. [6 marks]

Hint

Any word that ends in -ase is a type of enzyme.

7 a Explain why it is important to allow all the solutions to reach the temperature of the water bath before mixing the starch solution and amylase solution. [2 marks]

b It is best to monitor the temperature by placing a thermometer in one of the test tubes containing the starch solution, rather than placing it directly in the water bath.

Suggest why the thermometer is placed in the tube and not in the water bath. [2 marks]

8 Some bacteria live in volcanic vents where the temperature is 95°C. These bacteria can survive because their enzymes are specialised to work well at high temperatures.

On the axes below, sketch and label two lines to represent reactions controlled by:

- a human enzyme
- an enzyme from a volcanic vent bacterium. [4 marks]

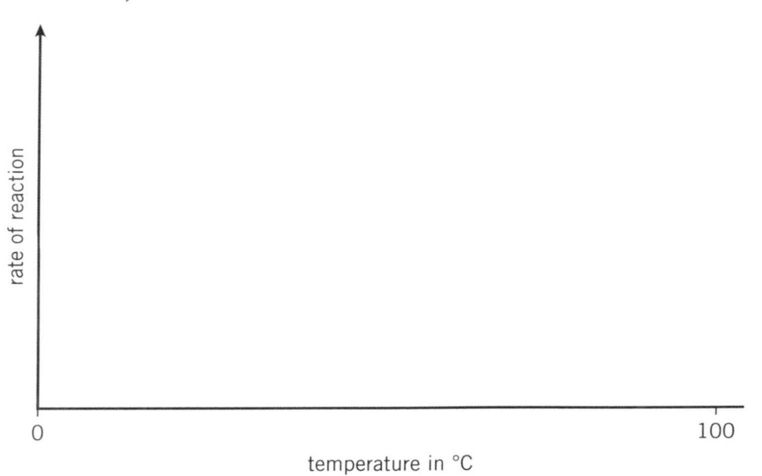

Exam Tip

When you're asked to sketch a graph, you need to include key details, for example, axes labels (if not already provided) and a line showing the correct shape of the graph, or the correct trend. You don't need to add number labels on the axes or to plot individual points unless you are specifically asked to do so.

9 The graph below shows the rate of reaction for an enzyme-controlled reaction at pH values from 0 to 4.

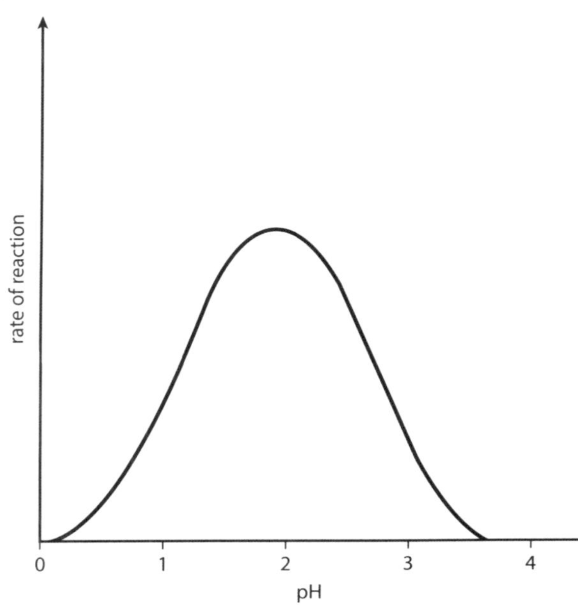

Explain the shape of the graph. [3 marks]

____

____

____

____

10 Three different classes carried out a version of this experiment.

Fresh amylase solution was made up for class A on Monday morning.

Class B used the same amylase solution at the end of Monday.

Fresh amylase solution was then made up for class C on Wednesday morning.

a Plot the results in the table below on the axes provided. [6 marks]

pH of solution	Average time for enzyme to break down substrate in s		
	Class A	**Class B**	**Class C**
4	43	72	35
5	25	52	19
6	57	198	37
7	170	340	120

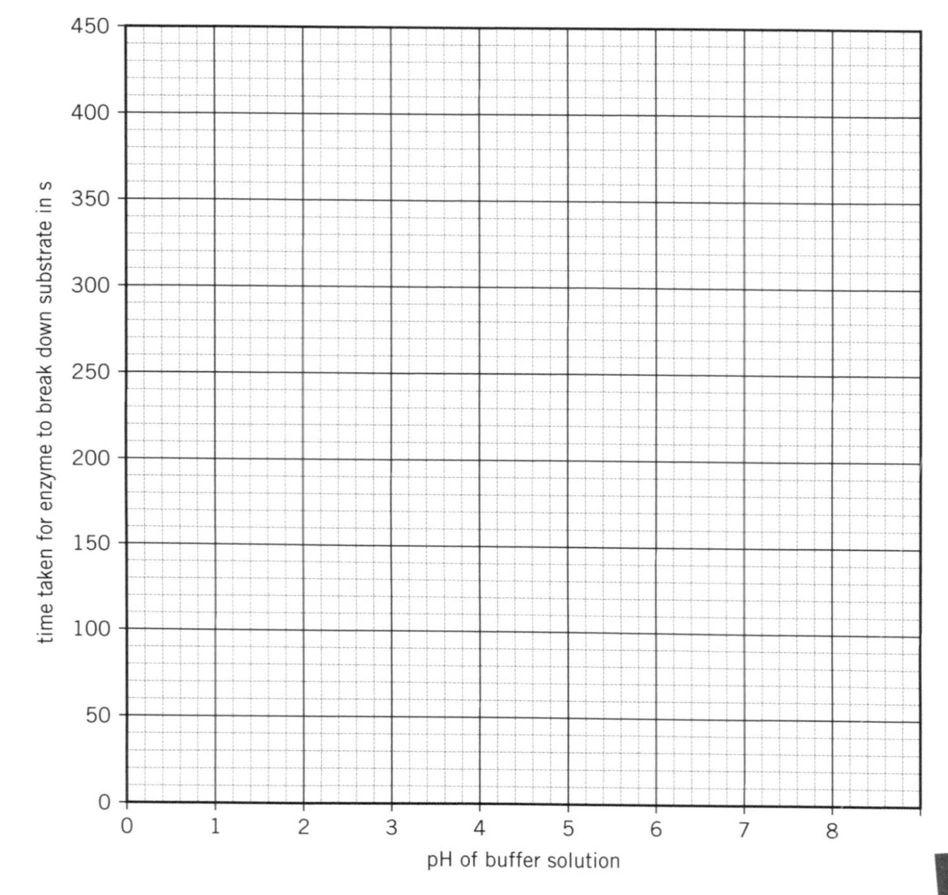

b Suggest reasons for the differences between the results of each class. [2 marks]

____

____

____

____

Hint

This graph may look unusual or upside-down, don't let that confuse you. Just read the axes carefully and think logically about what the data is saying.

c
- Estimate the optimal pH for the enzyme used.
- Give evidence from your graph to support your estimate. [2 marks]

11 A student carried out this experiment but found that the reaction happened so quickly that even at 10 seconds there was no colour change.

Suggest how the student could change their method to make the results easier to measure. [2 marks]

12 Describe why it is important to rinse the glass rod with water between each sample. [1 mark]

6 Photosynthesis

Investigate the effect of light intensity on the rate of photosynthesis.

Method

1. Cut a 10 cm piece of pondweed.
2. Place the piece of pondweed into a beaker of water, covered with an inverted filter funnel. Make sure the cut end of the pondweed is at the top.
3. Fill a measuring cylinder with water and carefully invert it over the top of the filter funnel.
4. Position a lamp exactly 100 cm from the pondweed. Switch the lamp on and leave it for two minutes to allow the pondweed to acclimatise.
5. Start a stopclock and record the number of bubbles produced in three minutes.
6. After 3 minutes, record the volume of gas that has been collected in the measuring cylinder.
7. Refill the measuring cylinder with water and repeat steps 4–6 for distances of 80 cm, 60 cm, 40 cm, and 20 cm.

Safety

- Wash hands after contact with pondweed and pond water.
- Lamp may get hot.
- Keep electrical equipment dry and do not handle if hands are wet.
- Dispose of the pondweed responsibly.

Equipment

- pondweed
- scissors
- lamp
- large beaker
- filter funnel
- 10 cm^3 measuring cylinder
- metre ruler
- stop clock

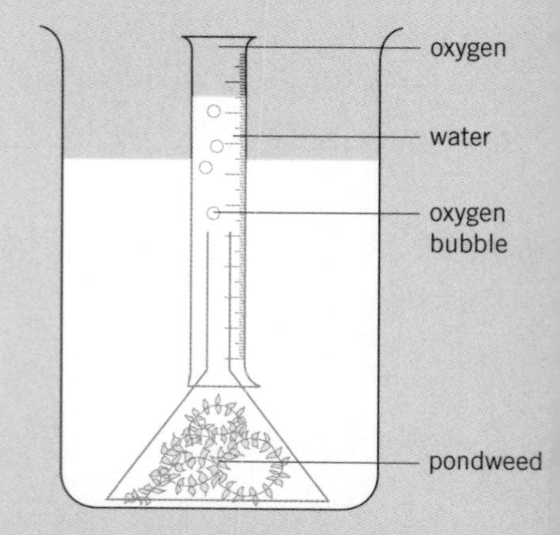

Remember

The skill being tested in this practical is your ability to accurately measure changes in the rate of photosynthesis in response to changes in the environment. You should be able to describe how to measure the rate of a reaction or biological process by collecting a gas that is produced.

Remember that you will probably have plotted a graph of the volume of gas collected (or number of bubbles) against the distance between the lamp and the pondweed (or light intensity). It is important to know what this graph looks like and to be able to explain its shape.

Exam Tip 

- A living plant is needed for this experiment to work properly. Plants only undergo photosynthesis in the right conditions, and a number of factors need to be controlled. The following questions will change some of those factors, some in an obvious way and some in a not so obvious way.
- You need to know the word equation for photosynthesis. You should also be able to recognise the chemical formulae for all the substances in the equation.

1 Identify the dependent variable in this experiment. [1 mark]

....

....

2 Draw one line from each compound to its chemical formula. [3 marks]

carbon dioxide	$C_6H_{12}O_6$
glucose	H_2O
oxygen	CO_2
water	O_2

3 Complete the word equation for photosynthesis. [4 marks]

.... + ... $\xrightarrow{\text{Light}}$ .... + ...

4 Identify the gas in the bubbles produced at the cut end of the pondweed.

[1 mark]

....

5 Two students carried out this practical with different light sources.

- Student A used a desk lamp.
- Student B used the light on their mobile phone.

Suggest how you would expect the students' results to differ and explain why.

[3 marks]

....

....

....

....

6 Two groups of students carried out the experiment in exactly the same way, their results are show below.

Group A

Distance to light source in cm	10	20
Number of gas bubbles	120	74
Volume of gas collected in cm^3	19	13

Group B

Distance to light source in cm	10	20
Number of gas bubbles	40	27
Volume of gas collected in cm^3	18	12

Compare the two sets of results. [4 marks]

Exam Tip

When you have 'compare' as the command word, you need to mention the similarities and the differences in your answer.

7 A student has samples of two different species of pondweed, shown below.

Explain whether it would be a fair test to compare rates of photosynthesis between these two samples of pondweed. [3 marks]

8 The following results were obtained after repeating the experiment. C omplete the table.

Distance from light source in cm	Number of gas bubbles released in one minute			
	Test 1	**Test 2**	**Test 3**	**Mean**
10	40	42	38	
20	27	25	68	
30	15	17	19	
40	11	12	9	

Exam Tip

Whenever you are asked to calculate the mean in a table of results, you should check for anomalous results first as there will often be one hiding in there. Do not include this result when working out the mean.

9 A student repeated the experiment twice in a day. The sample of pondweed was:

- tested at the start of the day
- left in water in the sun for the duration of the day
- retested at the end of the day.

Suggest why the volume of gas collected at the end of the day was much lower than the volume of gas collected at the start of the day. [4 marks]

10 Give the name of the structure in plant cells where photosynthesis takes place. [1 mark]

11 A student sets up their apparatus with air already trapped in the measuring cylinder before starting the experiment.

a Identify the type of error that the student has introduced. [1 mark]

Anomalous error ☐
Human error ☐
Random error ☐
Systematic error ☐

b Describe how the results would be adjusted to compensate for this error. [1 mark]

12 Suggest an alternative to a measuring cylinder that could be used to accurately measure the volume of gas collected. [1 mark]

____

____

13 A student investigates the effect of light intensity on the rate of photosynthesis. Their results are shown in the graph below.

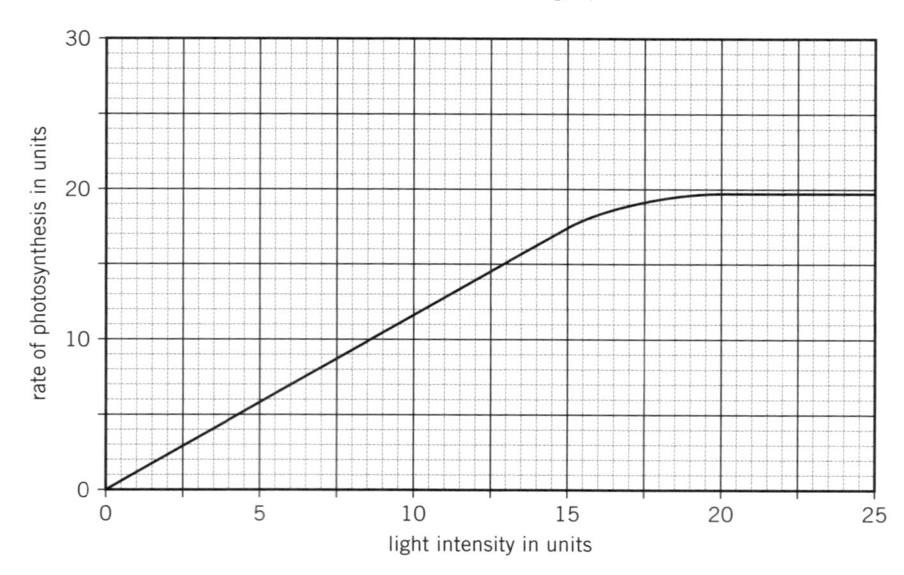

a Give the rate of photosynthesis when the light intensity is 12.5 units. [1 mark]

Rate of photosynthesis = ______________ units

b Increasing the light intensity beyond 20 units will not lead to a greater increase in the rate of photosynthesis.

Give evidence from the graph that supports this statement and explain why this is the case. [3 marks]

Evidence ______________________________

Explanation ______________________________

____

____

Exam Tip

Make sure you draw construction lines on the graph. These will help you work out the answer and will show the examiner you know what you're doing.

14 Describe possible sources of error in this experiment and suggest how the experiment could be adapted to reduce them. [6 marks]

15 (H) As the distance between the lamp and the pondweed increases, the light intensity decreases according to the inverse square law.

Explain the inverse square law. [4 marks]

Exam Tip

You might find it helpful to include a diagram in your answer – but don't spend too long drawing it!

7 Reaction time

Plan and carry out an experiment to investigate how one factor affects human reaction time.

Method

Human reaction times can be affected by several factors. You are going to choose one factor to investigate.

You will write down a hypothesis about how you think reaction time will change as you change your factor.

The next thing you will do is to test your hypothesis.

1. Working in pairs, one student holds a ruler vertically so that zero is at the bottom. The other student rests their hand on the edge of the bench and puts their thumb and fingers in a 'C' shape around the ruler, level with the zero marking.
2. Without warning, the first student drops the ruler and the second student has to catch it as quickly as they can.
3. Write down the number just above the second student's thumb. The lower the number, the faster the reaction time. Write the number in the results table.
4. Repeat steps 1–3 another four times.
5. Change the factor that you have decided to investigate (e.g., caffeine consumption, being distracted, gender) and repeat steps 1–4.
6. Use a conversion table to convert the results into reaction times.
7. Calculate mean values for each set of five results and plot your mean results on a bar chart.

Equipment

- metre ruler
- reaction time conversion table
- caffeinated soft drink (if caffeine consumption is the factor being changed)

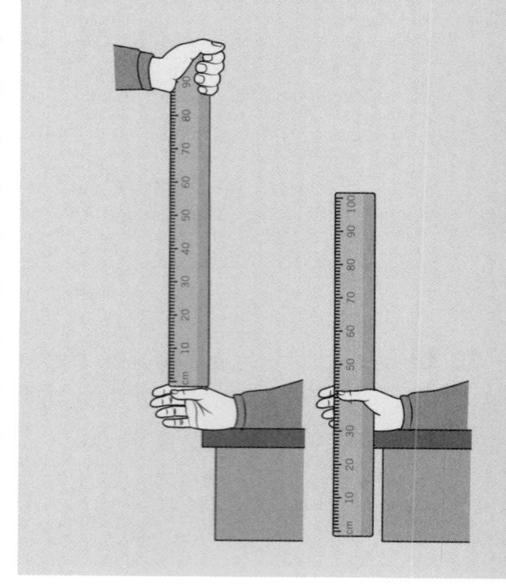

Safety

- Do not consume drinks in the laboratory.

Remember !

This practical tests your ability to plan an experiment and choose suitable variables to change in your investigation. You should be able to write a hypothesis predicting the effect that changing one variable will have on your results. Make sure you know what independent and dependent variables are.

Exam Tip 

You don't need to be able to calculate reaction time from the ruler measurement. In the practical you are given a table to look this data up, so make sure you know how to use the table.

1 **'Reaction time is affected by how tired a person is.'**

Describe how you would change the independent variable to test this hypothesis.

[3 marks]

___

___

___

___

___

___

___

Hint

The number of marks can give you a clue about how much detail to include. There are only three marks here so you are not going to be expected to re-write the whole method in detail.

Focus on the details of how you would change the factor.

2 A student wants to test the hypothesis:

'Reaction time is affected by taking illegal drugs.'

Explain whether this would be a suitable hypothesis to test.

[2 marks]

___

___

___

___

___

3 Draw one line from each factor to the effect it will have on reaction time.

[3 marks]

Factor being changed	Effect on reaction time
Caffeine	Decrease reaction time
Alcohol	Increase reaction time
Being distracted	No effect

Hint

Pay close attention to the wording of the effects and don't confuse reaction speed with reaction time, i.e., faster reactions = decrease in reaction time.

4 Three students recorded the following results for the ruler drop experiment.

Drop	Ruler measurement in cm		
	Student A	Student B	Student C
1	48.0	40	38.00
2	46.0	39	46.00
3	51.0	41	43.00
4	46.0	42	39.00
5	43.0	43	38.00
mean			
range			

a Complete the table by calculating the range and mean of each set of results. [5 marks]

b Give the letter of the student whose results are the most precise and give a reason for your answer. [2 marks]

Letter ______________

Reason ______________________________

____

c Use the following formula to calculate which student's results have the greatest percentage uncertainty. [4 marks]

$$\text{percentage uncertainty} = 100 \times \frac{(\text{range} \div 2)}{\text{mean}}$$

____

____

____

Student with greatest percentage uncertainty = ______________

d Explain the difference between precise results and repeatable results.

[2 marks]

____

____

____

____

5 Two students are investigating whether listening to music has an effect on reaction time.

a Explain why it is important that they use the same hand to catch the ruler each time. [2 marks]

____

____

____

Exam Tip

If you are being asked about why a variable should be kept the same, it is usually worth starting your answer with 'To make it a fair test'.

b State two other variables which should be kept the same in the students' experiment. [2 marks]

Variable 1 ______________________________

Variable 2 ______________________________

6 Describe what happens in a student's nervous system as they react to the falling ruler. [6 marks]

7 Two groups of students measured reaction times in different units.

One group recorded the data in mm and the other group recorded it in metres.

Give 0.28 m in mm. [1 mark]

Answer = __________ mm

8 Two students used an alternative method using a stopwatch.

Student A started the stopwatch as they released the ruler.

When they saw student B catch the ruler, student A stopped the stopwatch.

They then recorded the reaction time directly from the stopwatch.

Evaluate this method. [4 marks]

Exam Tip

When you see the command word 'evaluate', you need to give points for and against the method. You should then use these points to justify your opinion about whether the method is a good idea or not.

9 Two students have a hypothesis that:

'Reaction times will become faster with more practice.'

To test this hypothesis they repeat the ruler drop ten times to see if there is any change.

a Plot their results shown in the table on the axes provided. [4 marks]

Drop attempt	Reaction time in seconds
1	0.32
2	0.31
3	0.30
4	0.28
5	0.17
6	0.27
7	0.26
8	0.24
9	0.23
10	0.22

Exam Tip

When there is no scale provided, you will have to label your own axes. You will be given a sensible size of graph paper, so if your scale doesn't use most of it, check that you're going up in sensible amounts. If your scale uses less than half the space then you can double it!

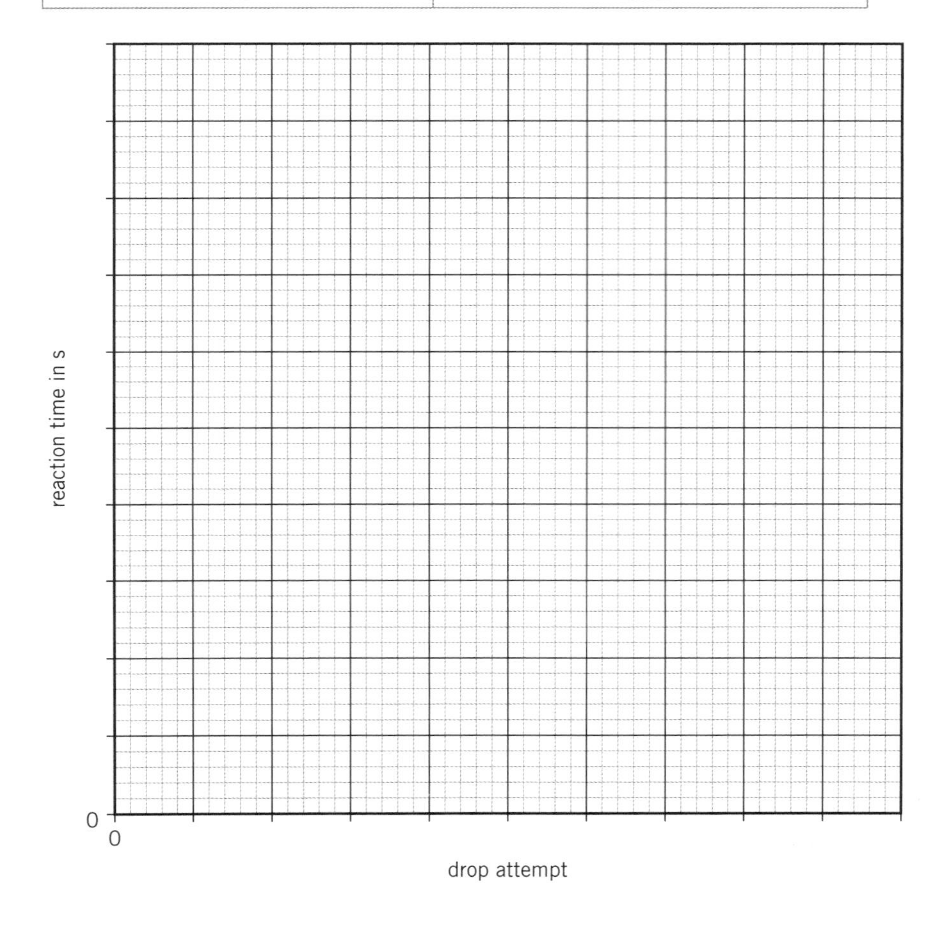

b Identify which result was an anomaly. [1 mark]

Anomalous drop number = ____________________

c Describe the pattern that can be seen in the results. [2 marks]

____

____

d Explain whether the results support the students' original hypothesis. [2 marks]

____

____

____

e The students then repeated the all ten drops while the 'catcher' was watching TV at the same time.

On your graph in part **a**, sketch a line of the results you would predict for this experiment. [2 marks]

8 Plant responses

Investigate the effect of light on the growth of newly germinated seedlings.

Method

1. You will be given three Petri dishes. Each will contain five newly geminated seedlings growing in a layer of cotton wool or compost.
2. Measure the length from the base of the shoot to the tip. Write your measurements in your results table.
3. Make accurate labelled drawings of the seedlings.
4. Using scissors cut out one side of the box.
5. Place one Petri dish:
   - inside the box (partial light)
   - next to the box (full light)
   - in a cupboard (darkness).
6. Put a lamp or light bank above the box and Petri dishes, making sure that some light is reaching the inside of the box.
7. You will need to add water during the experiment to make sure that the cotton wool or compost stays moist but not waterlogged.
8. Every day, for at least five days, measure the length from the base of the shoot to the tip. Write your measurements in your results tables (one for each light condition).
9. Make accurate labelled drawings of the seedlings.

Equipment

- newly germinated seedlings (e.g., mustard seeds)
- three Petri dishes
- cotton wool or compost
- water and pipette
- small cardboard box (e.g., a shoebox)
- scissors
- lamp or light bank
- ruler

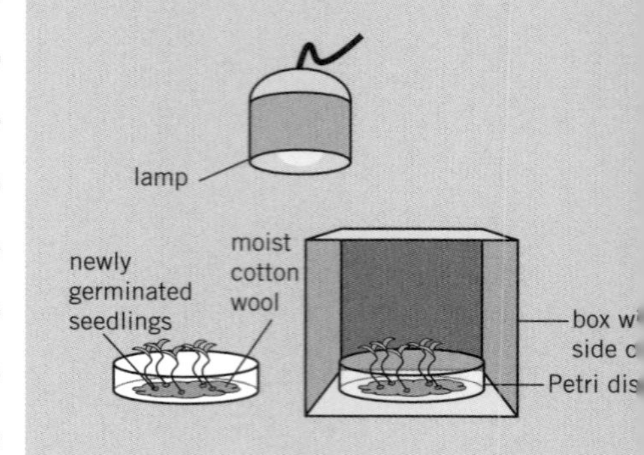

Safety

- Take care when using scissors.
- Wash hands after handling compost.
- Be aware of any students with relevant allergies.

Remember

The skills being tested in this practical are your ability to make and record measurements of length and time in order to measure biological changes.

It is an opportunity to practise recording observations as both quantitative measurements and as accurate, labelled biological drawings.

The plants you use in this experiment are living, so it is important to handle them with care.

Exam Tip 

You can be asked to apply any part of the maths you need to know to any part of the science you need to know, so don't be surprised if an unexpected maths question comes up.

1 Give the meaning of the term germination. [1 mark]

____

____

2 Give **three** variables that should be controlled in this experiment. [3 marks]

____

____

____

Exam Tip

Remember that to make a test fair, only one factor should change. Every other factor that could affect your results should be controlled.

3 Which process causes the seedlings' roots to grow down into the cotton wool?

Tick **one** box. [1 mark]

Gravitropism ☐

Gravity ☐

Photosynthesis ☐

Phototropism ☐

4 **a** White mustard or cress seeds are normally used for this experiment.

Suggest **two** reasons why these seeds are suitable for this experiment. [2 marks]

____

____

____

____

b Describe **one** problem that may be associated with taking measurements every day for at least five days. [1 mark]

____

____

c Explain why it is important that the seeds have all germinated before being placed in full light, partial light, or in darkness. [2 marks]

Hint

Think carefully about what you are investigating in this experiment and why using seeds that haven't germinated might affect your results.

5 A student grew eight seedlings growing in full light. They measured the height of the seedlings every day for five days.

a Complete the table below by calculating the mean seedling height for each row of data. [3 marks]

Time in days	Seedling height (full light) in cm								
	Seedling number								Mean
	1	2	3	4	5	6	7	8	
1	2	3	4	3	2	1	3	4	
2	3	5	5	4	2	3	4	5	
3	4	6	7	6	3	5	5	5	
4	5	7	9	8	3	7	6	6	
5	7	8	10	9	3	9	7	7	

b Calculate the rate of growth of seeding number **1** in millimetres per hour. [2 marks]

Answer = ______________ mm/h

The mean heights of seedlings grown in partial light and in darkness are given in the table below.

Time in days	Mean height of seedlings in cm	
	Seedlings in partial light	Seedlings in darkness
1	2.2	3.5
2	3.6	5.2
3	5.2	6.9
4	6.4	8.8
5	7.2	10.0

c Plot a graph of seedling height in cm against time in days on the graph paper provided. [6 marks]

Exam Tip

If you're asked to plot more than one set of data, make sure you label each line clearly.

d Suggest why the seedlings grew taller in darkness than in light and explain how they are able to do this in the absence of light. [3 marks]

6 A student investigated what happens if a seed is germinated lying in different positions. They observed that the shoot grows upwards and the roots grow downwards regardless of the orientation of the seed.

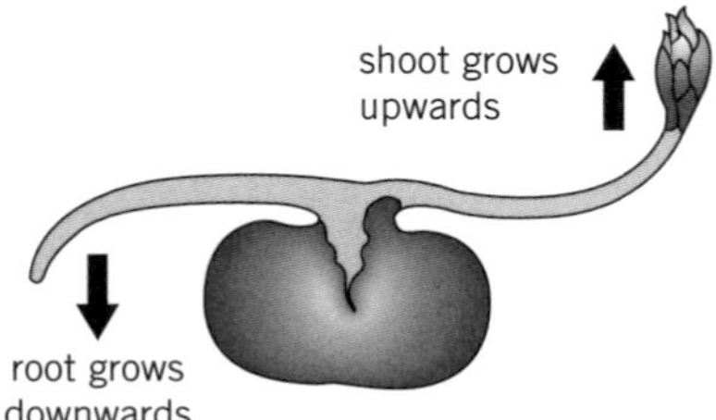

a Use your knowledge of auxins to explain the student's observations. [4 marks]

b (H) There are a number of other important plant hormones.

Draw one line from each hormone to its application. [3 marks]

Hormone	Application
Auxins	Increase the size of fruit
Ethene	Used as a weed killer spray
Giberellins	Control the ripening of fruit

7 A seedling was 4 cm at the beginning of the experiment. Its height increased by 265% over three days.

Calculate the new height of the seedling. [2 marks]

Height of seedling after three days = ____________ cm

8 A group of students wanted to investigate the effect of variables other than the presence or absence of light.

Describe a method they could use to investigate the effect of moisture levels on the growth on newly germinated seedlings. [6 marks]

Exam Tip

If you are asked to write a method or a plan, you can use a labelled diagram to help you describe the equipment. It only needs to be a labelled sketch so don't spend too long on it.

9 Field investigations

Investigate the population size of a plant species in a habitat.

Method

A Transect line

1. Stretch a 20 m tape measure from the base of a tree to an open area of ground.
2. Place the quadrat at exactly 2 m on the tape measure.
3. Count and record the number of plants of the species being investigated that fall within the quadrat.
4. Record the light intensity at this point on the transect line.
5. Repeat measurements every 2 m along the tape measure.

B Random sampling

1. Place two 20 m tape measures (labelled X and Y) at right angles to each other to form the sides of a 20 m^2 square area.
2. Put two sets of cards in a bag, each with the numbers 1 to 20 on them.
3. Pull two numbers out of the bag. The first number indicates how many metres along tape measure X you should move. The second tells you how far along tape measure Y to move.
4. Place the quadrat at these co-ordinates.
5. Count and record the number of plants of the species being investigated that fall within the quadrat.
6. Repeat steps 2–5 until you have sampled ten quadrats.

Safety

- Follow local rules on working in an outside environment and wash hands after the lesson.
- When any fieldwork is undertaken, work in groups and be aware of any hazards in that specific environment.
- Sensible footwear and clothing should be worn. If the weather is hot and sunny, sunscreen and hats are required.

Equipment

- two 20 m measuring tapes
- two sets of 20 cards, each numbered from 1 to 20
- 50 × 50 cm gridded quadrat frame
- notebook and pencil
- identification sheet
- optional equipment to measure abiotic factors, such as light meter, pH meter/ universal indicator paper, anemometer

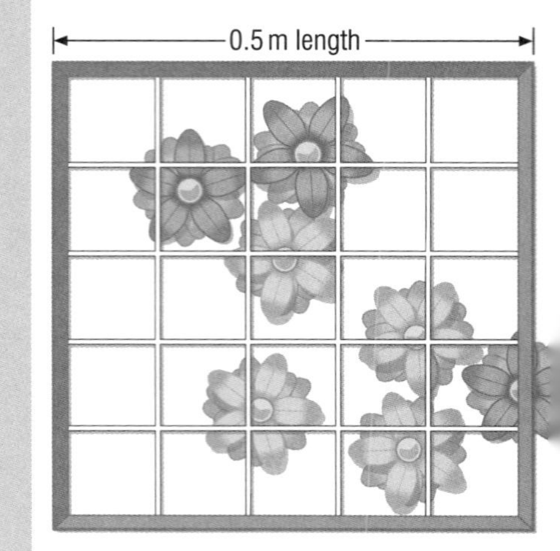

Remember

This practical tests your ability to accurately measure length and area and to apply appropriate sampling techniques in the field. There are two different methods of sampling with quadrats covered by this practical. It is important that you can describe how to do both of them and describe what the purpose of each method is.

Exam Tip

You should know how to calculate an estimate of the total population of a species using the equation:

$$\text{estimated population size} = \frac{\text{total area}}{\text{area sampled}} \times \text{number of individuals counted}$$

1 Why it is important to use a system to generate random coordinates instead of just choosing 'random' locations to place the quadrat.

Tick **one** box [1 mark]

Avoids unconscious bias ☐

Increases sample size ☐

It is faster ☐

Reduces errors ☐

2 Define 'habitat'. [1 mark]

....

....

3 A company wants to build a new headquarters on a piece of grassland that has an area of 21 000 m^2. They have been told they must carry out an environmental assessment before building begins.

The environmental assessment is carried out by randomly sampling ten 1 m^2 areas within the grassland.

a Suggest why it is important to investigate the species in the grassland, and to estimate their population sizes and distributions. [2 marks]

....

....

....

....

b Give an advantage of estimating population size instead of measuring the true population size in the grassland. [1 mark]

....

....

c Suggest an improvement to the method that would improve the accuracy of the population size estimate. [1 mark]

....

....

4 The graph below shows measurements of soil pH and the % grass coverage along a transect line in a garden. Soil pH has been plotted.

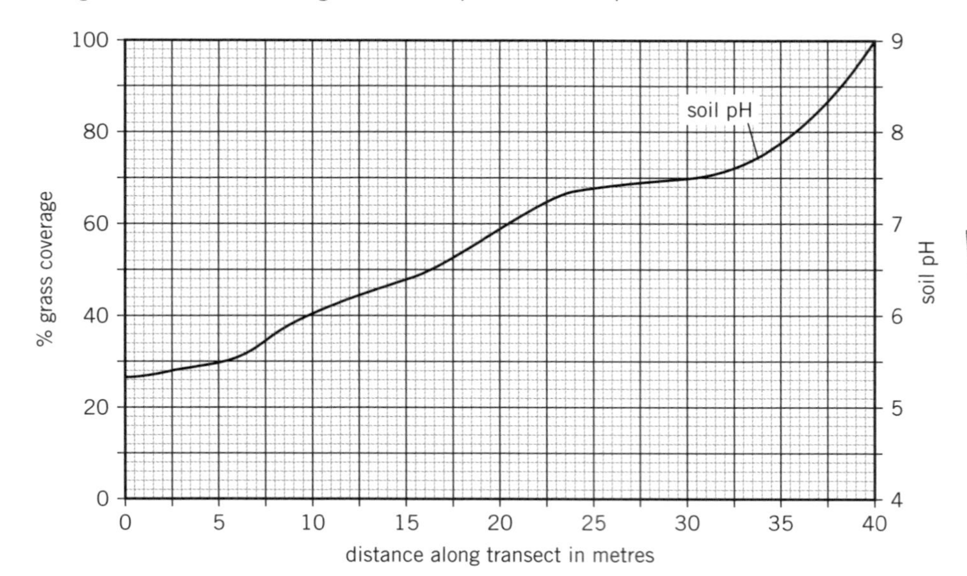

a Give the pH of the soil 20 metres along the transect line. [1 mark]

pH of soil = ________

b Plot the following data on the graph.

Distance along transect line in m	% grass coverage
5	50
10	85
15	95
20	90
25	35
30	40
35	30
40	15

c A gardening advice website says:

'Grass grows best in soil with a pH between 6 and 6.8.'

Explain whether the graph supports this statement. [2 marks]

____

____

____

____

Exam Tip

Be careful when graphs are shown with more than one y-axis. Make sure you read the correct axis. In this case, the soil pH is shown on the right.

5 Give 1 m^2 in cm^2. [1 mark]

____

____

1 m^2 = ________________ cm^2

Hint

A common mistake is to write 100 cm^2 – this is not the correct answer.

6 A student investigates the distribution of plantain weeds along a transect line and writes a list of factors they think might affect the distribution of the weeds.

Complete the table by sorting the student's list into biotic and abiotic factors.

[3 marks]

nutrient levels   light intensity   parasites   soil pH
primary consumers   temperature   wind

Biotic factors	Abiotic factors
competition from other plants	moisture levels

7 Describe a method to estimate the total population of ash trees in a 10 km^2 section of forest. [6 marks]

8 Random sampling and transect lines have different purposes.
Describe the differences in what the two methods are used to investigate.

[4 marks]

9 Calculate the area of a 25 cm × 25 cm quadrat in m^2. [2 marks]

Area = ______________ m^2

10 A student investigated the distribution of two plant species along a transect line. The transect started next to a wall and finished in the middle of a sunny school field.

Exam Tip

The command word 'compare' means you need to describe the similarities and/or differences between things. Don't just describe each of them on their own.

Quadrat number	Distance from wall in m	Number of plant A	Number of plant B
1	0	0	10
2	1	0	10
3	2	1	11
4	3	0	7
5	5	1	5
6	9	0	4
7	15	1	1
8	17	2	0
9	19	3	0
10	25	5	0

a Compare the results for the two plant species. [4 marks]

b Look again at the data in the table.

Suggest a change to the student's method that would have improved the data. [1 mark]

c A student found that plants near the wall had larger leaves than the ones in the school field.

Explain this finding. [2 marks]

11 A student used a 75 cm × 75 cm quadrat to randomly sample daisies in the school field. The field measured 30 m × 75 m.

Quadrat number	Number of daisies counted
1	12
2	21
3	13
4	89
5	24
6	19
7	10
8	29
9	21
10	16

a Calculate the mean and median number of daisies. [2 marks]

Hint

The median of an even set of numbers is the mean of the middle two numbers when the numbers are sorted into ascending order.

Mean = ________ daisies

Median = ________ daisies

b Use the data collected to estimate the population size. [5 marks]

Estimate of population size = ________ daisies

10 Decay

Measure the pH change of milk to investigate how temperature affects its rate of decay.

Method

1. Label a boiling tube 'milk' and add five drops of Cresol red solution, 5 cm^3 full-fat milk and 7 cm^3 sodium carbonate solution.
2. Label a second boiling tube 'lipase' and add 5 cm^3 lipase solution to it.
3. Place both tubes in a water bath set to 25 °C and place a thermometer in the 'milk' tube and wait until the contents of the tubes have reached the correct temperature.
4. Use a dropping pipette to add 1 cm^3 from the 'lipase' boiling tube to the 'milk' boiling tube and start the stopwatch at the same instant.
5. Use a stirring rod to gently stir the contents of the 'milk' boiling tube.
6. When the colour changes from purple to yellow, stop the stopwatch and record the time in a suitable results table.
7. Repeat the experiment at temperatures of 35 °C, 45 °C, 55 °C, and 65 °C.

Safety

- Do not drink the milk
- Wash your hands after the experiment
- Take care when using water baths at high temperatures

Equipment

- 250 cm^3 beakers
- boiling tubes
- Cresol red solution
- dropping pipettes
- full-fat milk
- lipase solution
- 10 cm^3 measuring cylinders
- sodium carbonate solution
- stirring rod
- stopwatch
- thermometer
- water bath

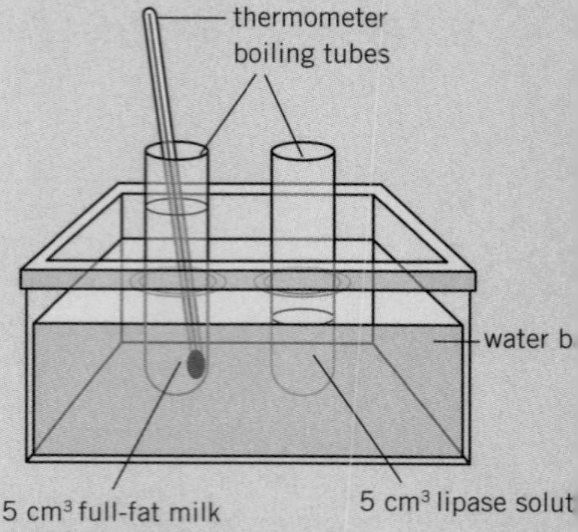

Remember

In this practical you plan and carry out an experiment to find out how the rate of decay of fresh milk changes as you change the temperature. The practical tests your ability to observe biological changes by making accurate measurements of time and temperature. You should be able to describe how an indicator solution can be used to measure the rate of a reaction.

Exam Tip

Milk decays very slowly over time producing lactic acid which causes the pH to decrease. In this experiment lipase is added to model this process allowing you to observe a change in pH within lesson time. The pH change in this experiment is caused by a different substance.

Remember that this is only a model of the real process.

1 Draw **one** line from each solution to its function in this experiment. [4 marks]

Solution	**Function**
Cresol red	Substance that is decaying
Full-fat milk	Indicator solution
Sodium carbonate	Enzyme to speed up decay process
Lipase solution	pH buffer

2 List three variables which should be controlled in this experiment. [3 marks]

____

____

____

____

____

____

Exam Tip

Lots of students talk about 'amounts' of things. This word is not specific enough!.

Use words like volume, mass, or concentration to move your answers up to the next level.

3 List **three** things that can affect the rate of decay of substances in the milk in this experiment. [3 marks]

____

____

____

Hint

Questions 2 and 3 are more similar than they may seem at first.

4 Suggest an alternative to a pipette that could be used to measure 5 cm^3 of milk. [1 mark]

____

____

5 A group of students followed the method described, but decided to start at 10 °C and then increase the temperature every time by 2 °C until they reached 60 °C.

a Explain whether the results of this experiment are qualitative or quantitative. [2 marks]

____

____

b Evaluate the design of the students' experiment. [3 marks]

____

____

____

____

____

____

Exam Tip

Remember that when you are asked to 'evaluate' you should use the information given in the question and your own knowledge to give evidence for and against something.

6 A student carried out the experiment described in the method. Their results are represented in the graph below.

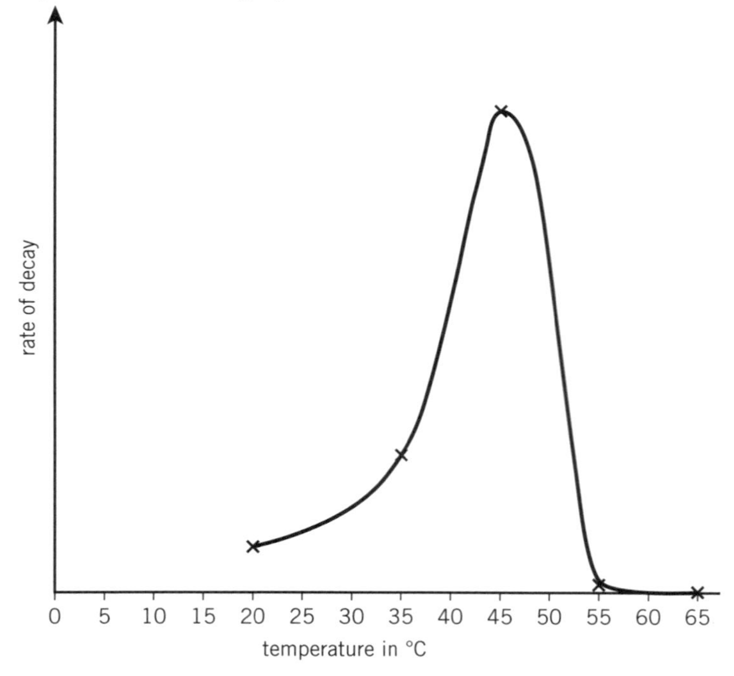

a Which substance in milk is broken down by lipase?

Tick **one** box. [1 mark]

Carbohydrate ☐

Fat ☐

Protein ☐

Starch ☐

b Compare and explain the results at 20 °C and at 65 °C. [4 marks]

____

____

____

____

____

____

c Use the graph to estimate the optimum temperature of the enzyme used in this experiment. [1 mark]

Optimum temperature = ____________ °C

d The decay of milk is too slow to observe in a single lesson as it will take several days for the pH to change due to the increase of lactic acid in the milk.

Lipase is added to model the decay of milk in a much shorter time, allowing the practical to be carried in a single lesson.

i Name the substance produced that causes the pH to change when lipase is added. [1 mark]

____

ii Explain how you would expect the results to differ if the decay of milk was observed without the addition of lipase. [4 marks]

____

____

____

____

____

____

____

____

Exam Tip

This question might seem daunting at first. Take it bit by bit, describing and explaining each difference you would expect to see in the results.

7 Two students carry out the experiment at different temperatures and shared their results in order to gather more data. One of the two students is worried that they are not stopping the stopwatch at the same point in the colour change.

Describe how the experiment could be changed to make sure the end-point is always consistent. [2 marks]

____

____

____

____

8 A student carried out a similar experiment, but used a pH probe to measure the change in pH over time. The student's data is shown below.

Time in minutes	pH
0	6.8
10	6.4
20	6.0
30	5.6

a On the axes provided below:

- plot the student's data
- draw a line of best fit. [4 marks]

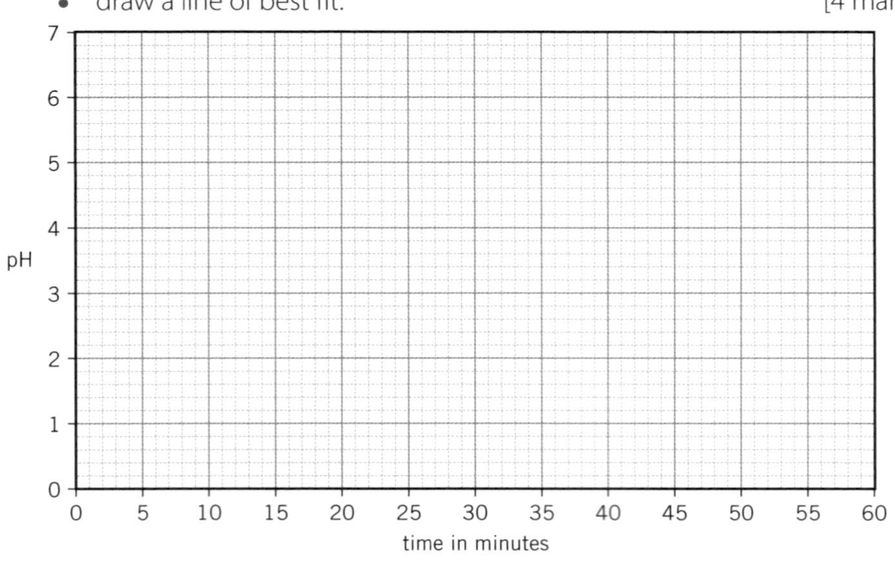

b Use your graph to predict the pH of the milk after:

i 25 minutes [1 mark]

Answer = ____________

ii 40 minutes [1 mark]

Answer = ____________

iii 60 minutes [1 mark]

Answer = ____________

c **i** Suggest which of the predictions you made in part **b** you would be **most** confident about and give a reason why. [2 marks]

____

____

____

ii Suggest which of the predcitions you made in part **b** you would be **least** confident about and give a reason why. [2 marks]

____

____

____

9 Describe why is it important to stir the mixture after the lipase solution is added. [1 mark]

____

____

10 A student carried out the experiment described in the method and recorded the following data.

Temperature of milk in °C	Time taken for solution to turn from purple to yellow in seconds
15	367
25	178
35	29
45	34
55	67

a Calculate the range of the results. [1 mark]

____

____

Range = __________ seconds

b Give 367 s in minutes and seconds. [1 mark]

____

____

Answer = __________ minutes __________ seconds

c The students decide state that their best estimate of the optimum temperature is 35 °C.

Describe how the students could modify their method to find a better estimate of the optimum temperature. [3 marks]

____

____

____

____

____

11 Describe one possible source of error in the experiment described in the method and describe how this would affect the results. [2 marks]

____

____

____

____

Notes – 1 Microscopy

Make notes on the practical you carried out.

Hint

Make notes on:
- your method
- safety precautions
- sources of error
- possible improvements
- the function of equipment
- how to make accurate measurements.

Tip

You could use this space to:
- write the equation linking magnification, image size, and object size
- draw a labelled diagram of a light microscope.

Notes – 2 Microbiology

Make notes on the practical you carried out.

Hint

Make notes on:
- your method
- safety precautions
- sources of error
- possible improvements
- the function of equipment
- how to make accurate measurements.

Tip

You could use this space to:
- write the formula you would use to calculate the area of a clear zone
- sketch a graph of number of bacteria against time.

Notes – 3 Osmosis

Make notes on the practical you carried out.

Hint

Make notes on:
- your method
- safety precautions
- sources of error
- possible improvements
- the function of equipment
- how to make accurate measurements.

Tip

You could use this space to:
- write the equation for % change in mass
- sketch a graph of percentage change in mass against concentration of sugar solution.

Notes – 4 Food tests

Make notes on the practical you carried out.

Hint

Make notes on:
- your method
- safety precautions
- sources of error
- possible improvements
- the function of equipment
- how to make accurate measurements.

Tip

You could use this space to write what you would expect to observe in a positive test result for each of the food tests you carried out.

Notes – 5 Enzymes

Make notes on the practical you carried out.

Hint

Make notes on:
- your method
- safety precautions
- sources of error
- possible improvements
- the function of equipment
- how to make accurate measurements.

Tip

You could use this space to draw a sketch graph of:
- reaction rate against pH for an enzyme-controlled reaction
- reaction rate against temperature for an enzyme-controlled reaction.

Notes – 6 Photosynthesis

Make notes on the practical you carried out.

Hint

Make notes on:
- your method
- safety precautions
- sources of error
- possible improvements
- the function of equipment
- how to make accurate measurements.

Tip

You could use this space to:
- draw a labelled diagram of the equipment you used
- sketch a graph of rate of photosynthesis against light intensity.

Notes – 7 Reaction time

Make notes on the practical you carried out.

Hint

Make notes on:
- your method
- safety precautions
- sources of error
- possible improvements
- the function of equipment
- how to make accurate measurements.

Tip

You could use this space to sketch a bar chart of reaction time against caffeine consumption ('before caffeine'/'after caffeine')

Notes – 8 Plant responses

Make notes on the practical you carried out.

Hint

Make notes on:
- your method
- safety precautions
- sources of error
- possible improvements
- the function of equipment
- how to make accurate measurements.

Tip

You could use this space to:
- sketch a bar chart of seedling height against light conditions (light/partial light/dark)
- draw a labelled diagram of the equipment you used to investigate the effect of light on newly germinated seedlings.

Notes – 9 Field investigations

Make notes on the practical you carried out.

Hint

Make notes on:
- your method
- safety precautions
- sources of error
- possible improvements
- the function of equipment
- how to make accurate measurements.

Tip

You could use this space to:
- write the equation for estimating population size from a sample
- sketch a graph of your results.

Notes – 10 Decay

Make notes on the practical you carried out.

Hint

Make notes on:
- your method
- safety precautions
- sources of error
- possible improvements
- the function of equipment
- how to make accurate measurements.

Tip

You could use this space to sketch a graph of the rate of decay against temperature.

Notes

Notes

Notes

Notes

Notes

Notes

Great Clarendon Street, Oxford, OX2 6DP, United Kingdom

Oxford University Press is a department of the University of Oxford. It furthers the University's objective of excellence in research, scholarship, and education by publishing worldwide. Oxford is a registered trade mark of Oxford University Press in the UK and in certain other countries

First published in 2019

British Library Cataloguing in Publication Data
Data available

978 0 19 844493 0

10 9 8 7 6 5 4 3 2 1

Paper used in the production of this book is a natural, recyclable product made from wood grown in sustainable forests.
The manufacturing process conforms to the environmental regulations of the country of origin.

Printed in Great Britain by Bell and Bain Ltd. Glasgow

Acknowledgements

COVER: CHARLOTTE RAYMOND/SCIENCE PHOTO LIBRARY

Artwork by Aptara Inc.